DISCARDED
Goshen Public Library

FARM ANIMALS

James A. Corrick

GALLERY BOOKS
An Imprint of W. H. Smith Publishers Inc.
112 Madison Avenue
New York City 10016

This edition first published in U.S.
in 1990 by Gallery Books,
an imprint of W.H. Smith Publishers, Inc.
112 Madison Avenue, New York, New York 10016

ISBN 0-8317-9588-3

Printed and bound in Spain

For rights information about the photographs in
this book please contact:

The Image Bank
111 Fifth Avenue, New York, NY 10003

Producer: Solomon M. Skolnick
Author: James A. Corrick
Design Concept: Lesley Ehlers
Designer: Ann-Louise Lipman
Editor: Terri L. Hardin
Production: Valerie Zars
Photo Researcher: Edward Douglas
Assistant Photo Researcher: Robert Hale
Editorial Assistant: Carol Raguso

***Title page:* Cattle on a California ranch exhibit the characteristically square build of good beef animals. *Opposite:* A French cockerel rooster has his comb raised in alertness.**

A rooster crowing is a common barnyard sight. This Bantam rooster is announcing his presence to any would-be rivals. *Below:* The large-scale raising of chickens did not begin until the middle of the nineteenth century. Prior to that time, farmers kept a few chickens for their own use.

Farming and the keeping of farm animals marked an important change in human culture and society. Before farming, people lived in small nomadic groups, hunting animals and gathering whatever wild, edible plants could be found. Farming was the beginning of settled humanity; when people no longer needed to move from one locale to another, much larger human communities became possible. The eventual result was the birth of cities.

The keeping of farm animals also led, surprisingly, to the development of writing. All early writings are farm records, detailing the number of animals – goats, pigs, sheep, and cattle – that were bought, kept, or sold.

The change from hunting to farming was rapid. About 15,000 years ago, all humans were hunter-gatherers, but within a 5,000-year span, almost every culture was farm-based. And in some places in the world, farming and herding are still practiced virtually the same way as they have been done for hundreds, maybe thousands, of years.

Top to bottom: **Chicken farming is a very efficient business – one person can care for as many as 25,000 to 50,000 birds. Unlike many hens which spend their entire lives in hatcheries, eating processed feed, these Rhode Island red hens forage freely. These chicks have found shelter under the wings of their mother.**

Chicks will increase their body weight by 43 times over an eight-week period. *Below:* On this farm in China, farmers use baskets with metal caps as incubators to keep young chicks warm. The loose weave of the basket and the holes in the cap allow sufficient ventilation. *Opposite:* Some farmers, such as the owner of this farm near Frastanz, Austria, keep only enough chickens and turkeys for their own provision.

A farm provided its owners with a constant supply of meat and vegetables, whereas hunter-gatherers ate only when they found food. Farm animals became living larders and have remained so to the present time.

Farmers have attempted to make almost every animal, no matter how unlikely, into livestock. They have tried – and failed – to tame the giraffe, hippopotamus, and Pantagonian sloth. They have even tried cultivating the sponge and the ailanthus moth (which, with a five-inch wingspan, is about the size of a bat). Only a few species, however, proved successful as farm animals: ones that live easily and breed freely in captivity; unsuccessful ones die quickly.

Preceding page: This American turkey on a farm near Weston, Vermont, is not descended from the native North American turkey, but from a Mexican bird. *This page:* Most turkeys, such as these, will be sold for holiday dinners. *Below:* This male or tom turkey, with his fully extended fan-shaped tail, hopes to interest the two hens in the background in mating.

Turning wild pigs into the 300-plus breeds of present-day domestic hogs, for example, took a long time and plenty of work. Our ancestors decided what they wanted then bred the selected individuals. Centuries of experimentation and dead ends finally produced the myriad farm-animal breeds that are with us today.

The breeds of any particular animal are all part of the same species, but each breed has a set of characteristics distinguishing it from the rest of the animals of that species. Hereford, Indian Brahman, and Holstein are all separate breeds, but they are also all cattle. Two pure-blooded parents of the same breed will produce offspring that are members of that breed. Mate two purebred Holsteins, and you produce more Holsteins.

No one knows for certain which species became the first farm animal, but some archaeologists believe it was the reindeer. In the ice-age Europe of 20,000 years ago, the herds were larger, serving as prey to a number of predators; later, they served as prey to humans. What would be more logical than to begin taming such animals so as to

Top to bottom: **These baby ducks in Suzhou, China, will lose their yellow dow in a few weeks as they begin to grow feathers. Geese, although an important farm bird in central Europe, are raised only on a small scale in the rest of the world. These ducks and geese on an Amish farm pond near Lititz, Pennsylvani mingle freely.** ***Opposite:*** **Unlike chickens and turkeys, ducks are aquatic and must have water if they are to grow well.**

JOHN DE

Preceding page: **A pig farmer surveys his stock of Duroc-Jerseys, recognizable by the reddish tint of their coats.** ***This page:*** **Pigs have no sweat glands; they must cool down by partially submerging themselves in mud or water.**

provide a constant source of food and clothing? Reindeer are still an important domesticated species in Lapland, a region of northern Norway and Sweden.

Other archaeologists, however, claim the evidence for early reindeer domestication is missing. More likely, they argue, humans merely continued the same pattern of following and hunting reindeer for thousands of years, and it was not until long after the arrival of the dog that they domesticated the herds

The case for sheep as the first farm animal is sounder. Archaeologists have found solid evidence in northern Iraq that domestic sheep grazed that area as far back as 9000 B.C. Sheep would have been a particularly good choice for a herd animal in this area since they can survive on sparse vegetation and little water. The sheep's hardiness makes it a good source of meat and clothing fiber.

Sheep are ruminants, or cud-chewers, just like cattle; their four-chambered stomachs contain bacteria, helping them t digest grass. Both wild males an females have two hollow horns, which they do not shed. The horns of the male are often elaborate spirals, while those of the female are shorter and only slightly curved. Many commercial sheep have been bred to be hornless.

These sleeping piglets will quickly grow until they reach their adult weight of 200 pounds. *Below:* Sows generally have seven to eight piglets in each litter and can have up to two litters a year.

hese pigs live in a large building that can ld up to a thousand of them. Having elter makes them healthier and less one to disease than pigs in the open. *ight:* Pigs are intelligent animals and, e puppies and kittens, the young ones ay. This piglet is playing hide-and-seek th its littermates.

Sheep generally winter in ocks of 1,500 at lower altitudes. hey spend the summer, under e supervision of one or more epherds, in high-altitude astures, which can be as much 300 miles from their winter uarters.

Europeans used the ancestors of these pigs, found in China, to improve Western breeds. *Below:* The white pig is a Yorkshire – probably the most common breed of swine in the world. The red ones are Duroc-Jerseys. *Opposite:* This hog is tagged as part of a state fair competition. Such competitions play an important role in breeding selection.

Over 200 breeds of sheep exist today. Some, such as the large, white-faced British Cotswold, are long-wool sheep, while others, like the smaller Cheviot, with its woolless head (developed first in Scotland), are medium-wool sheep. The third general class of sheep is the fine wool. The most famous fine-wool breed is the merino, originally a Spanish strain, whose face seems to peek out of the mass of wool covering its head. The merino is also the ancestor of the larger, less wrinkled French Rambouillet. Together, the fine-wool breeds account for 50 percent of all the world's sheep.

The coat of the long-wool breeds is coarse with long fibers; since the market for long-fibered wool is small, these breeds are raised mostly for their meat. Medium- and fine-wool sheep have much softer coats with increasingly thinner fibers. The majority of the commercial wool of the world comes from these latter two breeds, although recent breeding experiments have produced fine wools, such as the Corriedale and Romendale that are good meat suppliers.

A few breeds (known as fat-tailed sheep because they store fat in their tails and rumps) are kept for milk. These are popular

Sheep, which outnumber the human population of Australia by eight to one, are an important part of the economy.

Sheep, such as these in Somerset, England, are still an important source of meat for Great Britain. *Opposite:* Shearing a sheep of its wool does not harm the animal, and it is never done until the worst of the winter weather has passed.

Preceding page: Each of the sheep on this Australian farm produces over eight pounds of wool a year. *This page:* Sheep dog trials in New Zealand (left), as they do elsewhere, test those dogs' ability. A good sheep dog (bottom) keeps his flock together with speed; barking would panic the sheep into scattering.

This shepherd in the Sierra Nevada region of Spain oversees the grazing of his sheep flock. Sheepherding has been practiced in this area for several thousand years. *Opposite:* This shepherd spends several weeks alone with his flock. Each day, he and his dog make regular rounds to check each grazing area of the flock.

Two fluffy sheep display their heavy winter coats. *Below:* These Italian sheep are a long-wool breed, raised for meat.

in the arid regions of Africa, the Middle East, and Asia, particularly the Awassi and the Karamo breeds. In other parts of the world, small flocks of sheep are used as scavengers to clear away weeds and brush.

Most sheep today are crosse between two or more breeds. Farmers favor long-wool males (or rams) and fine-wool females (or ewes) to produce lambs for market. One of the most popula breeds – by itself and as a crossbreeder – is the fine-wool merino. The medieval Spanish attempted to monopolize this sturdy sheep because of its superior wool, but the merino soon spread over Europe.

Not long after the first domestic sheep came the first tame goats. Goats are related to sheep but are smaller, with horn that do not curve and tails that point up rather than down. Like sheep, they are ruminants. Male goats are "rams" or "billys," while females are "does or nannys." Young goats are "kids."

Most goats are kept for their milk, which is actually easier for infants, invalids, and allergy sufferers to digest than cow milk. One or two goats can not only provide an entire family with milk, but they require less grazing land than even a single cow. Goats are also important for their meat, particularly in Mediterranean countries such as Italy and Greece.

Some of the most important odern breeds of goats are lpine, Nubian, Angora, and ashmir. The prick-eared lpine or Swiss goat is found roughout Europe and the nited States. It is an excellent ilk producer, and farmers ossbreed it with other varieties f goat in order to increase their ilk yields. The Nubian or astern goat spread from Egypt cross North Africa and the Iiddle East. A good milk roducer like the Swiss, it has ng, drooping ears and a large ose. Angora goats have long, lklike hair – called *mohair* – and ne smaller Kashmir goats, native India, produce cashmere wool at is used in shawls and veaters, among other garments.

Not long after the goat, rmers domesticated cattle. The rst tamed herds appeared in urope and India almost simul- neously some 8,500 years ago. attle provide a variety of roducts, including leather, glue, nd gelatin. Their most impor- nt products, of course, are beef nd milk. Most modern cattle reeds supply either milk or neat, but rarely both.

op to bottom: These merino sheep of ustralia are highly prized for their fine ool. This sheep, its winter wool still nsheared, rests among spring wildflowers ear Corfu, Greece. These Icelandic ieep resemble their wild cousins more an they do other domestic sheep.

Cattle may have been farm animals for over 10,000 years, but most modern cattle breeds are only two centuries old. One of the oldest of the modern breeds of beef cattle is the Hereford. British farmers developed this breed with its distinctive red and white markings and short, compact body in the middle of the eighteenth century. In the early nineteenth, this breed came to the United States; it is now the most popular strain of beef cattle in America.

Preceding page: **This ram on a farm in the Outer Hebrides, Scotland, has a set of spiraled horns.** ***This page:*** **Sheep are amazingly fecund. A female sheep or ewe (top) can have as many as three lambs a year for up to 20 years! Most sheep are a cross between two or three different breeds.**

Preceding page: **Goats, like cattle and sheep, are ruminants or cud-chewers. Bacteria in their multi-chambered stomachs helps them digest the grass they eat.** ***This page:*** **Most goats (top) are raised for their milk. One or two goats, with less grazing land than a single cow, can supply enough milk for an entire family. Goats (bottom) usually have two kids at a time.**

The Hereford, however, was nitially unable to survive in the American Southwest, where fever icks and various warm-weather liseases killed it. Nineteenth-entury Southwestern cattle anchers found the Indian Brahman was tick-resistant and, fter crossbreeding it with the Durham, they produced the onghorn or Santa Gertrudis reed. This breed dominated eef ranching from Texas to outhern California until modern eterinary medicine solved the ick problem.

While cattle ranching spread ver the western United States, airy farming came to dominate nuch of the American Midwest, articularly Wisconsin and Minnesota. The three major airy breeds are the black-and-hite Holstein, the fawn-and-hite Guernsey, and the fawn ersey. Like all dairy cows, these reeds are more angular and less quare than beef cattle. The olstein is the largest of the ree breeds and gives the most ilk. Its milk has the least butter-t; Jersey milk, by comparison, s the most. Farmers normally e Jersey milk to make cheese.

With modern milking machines, dairy farms such as these can have several hundred head of cattle.

Holstein cattle, a breed developed in the Netherlands, produce more milk than any other kind of dairy cows.

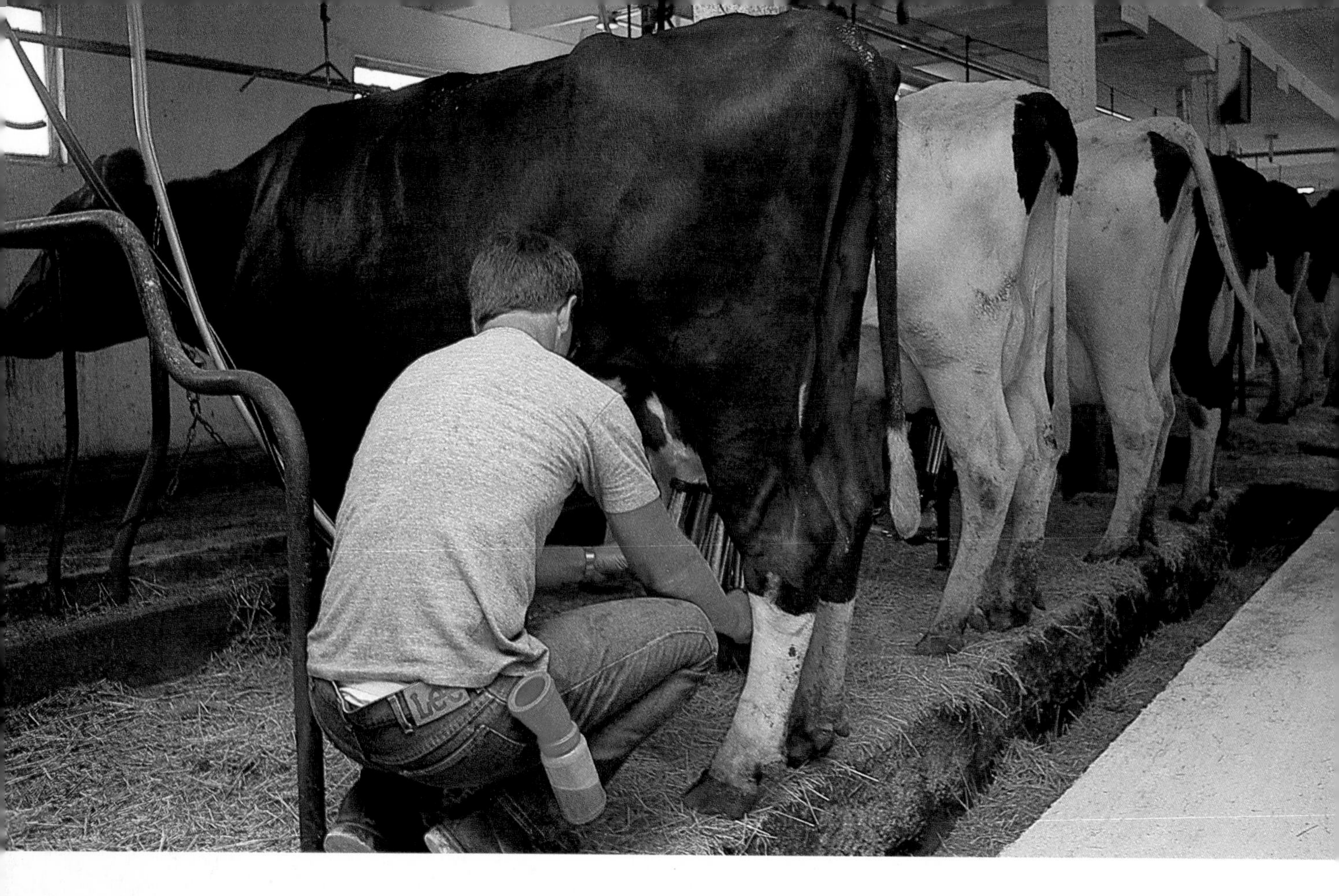

Dairy farmers must milk their entire herd twice a day. *Below:* Many ranches and farms have one or two dairy cows just for their needs, but wi have a milking machine. Such a small number of cattle would not be enough to justify the expense.

The Jersey, which originated on the Isle of Jersey in the English Channel, is the smallest of all dairy cattle.

The invention of the milking machine freed dairy farmers from the time-consuming and laborious process of hand milking. Dairy farms are now virtual factories that mass-produce milk, since a farmer with only a few helpers can milk hundreds of cows with these machines.

For meat production, no farm animal equals the pig. The pig, or swine, quickly and efficiently turns food into meat. It is also an omnivore, eating, like humans, both vegetables and meat (although most commercial hogs are fed on corn).

Pigs are also easily raised in large buildings, called nurseries or farrowing houses. Applying mass-production techniques to pigs has eliminated many of the problems with disease and parasites that have plagued swine farmers for thousands of years.

Top to bottom: **This Holstein has the breed's characteristic white and black markings. During the summer, Swiss cattle graze in high mountain pastures. These Holsteins find most of their nutritional needs in the grass of this pasture near their barn. *Opposite:* This cow on a farm near Hegeve, France, has a belt with a bell buckled around its neck. Listening for these bells, farmers can trac the location of their free-grazing herd.**

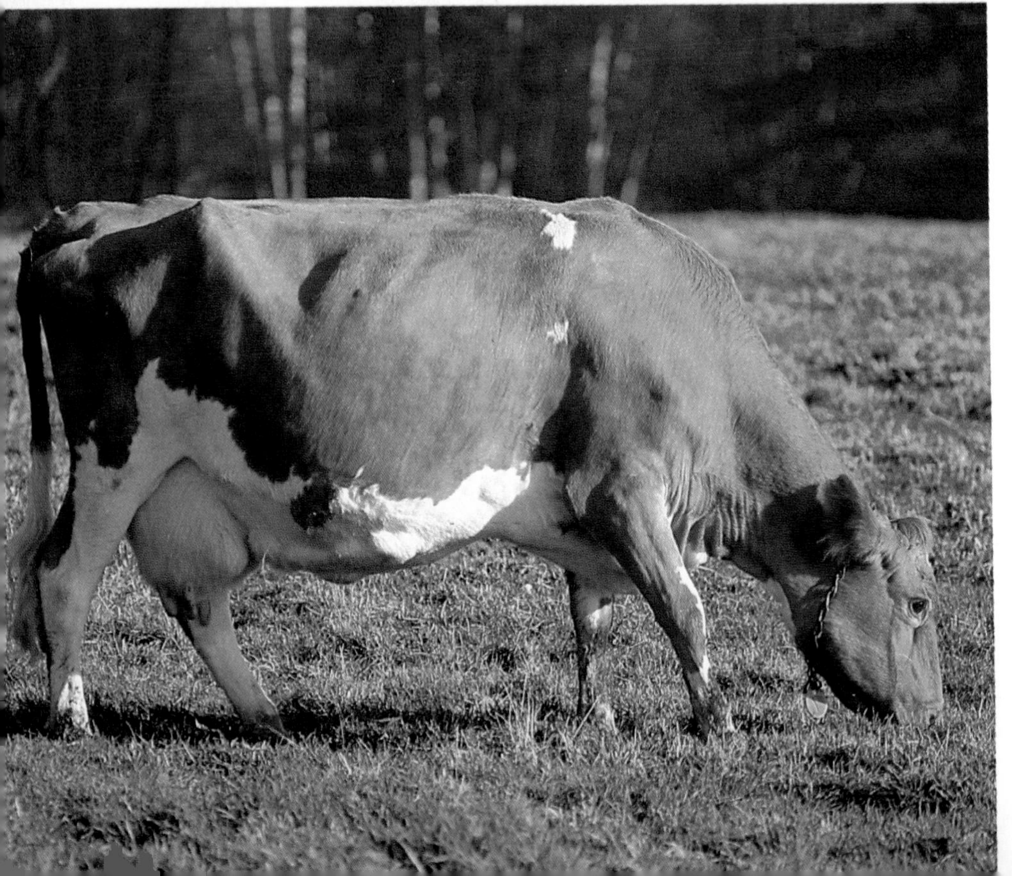

The pig is native to Asia, from where, by various human migrations, it was spread to Europe and Africa. The Spanish introduced the pig into the Americas in the fifteenth and sixteenth centuries; and all wild pigs on these two continents are descended from escaped European pigs. New World piglike animals are not actually pigs, although the javelina of the U.S. Southwest fills the same environmental niche occupied by the pig in Asia.

A pig is any swine that weighs less than 180 pounds; a hog any that weighs more. Up to the middle of the twentieth century, 220-pound hogs were routinely raised for lard. Now, most swine are smaller and leaner, being raised for bacon and pork. Denmark's Landrace breed has been famous for its bacon since the last century, and the breed is considered of such value that the Danish government now refuses to allow the export of any Landrace swine.

The Hereford, such as the one pictured here, is one of the oldest breeds of beef cattle. *Left:* Like the Jersey, this Guernsey dairy cow is a descendent of cattle from an island in the English Channel.

eifers, such as this Jersey, are young
male cattle halfway between being calves
d mature cows. *Right:* Farmers and
nchers breed beef cattle (such as this
e in Scotland) to be heavier and have
more square frame than dairy cattle.
verleaf: The Highland breed of beef
ttle is very hardy and has a heavy coat
r protection from fierce winters.

Preceding page: The King Ranch in Texas developed the Santa Gertrudis or longhorn breed of beef cattle. For nearly a century, ranchers in the American Southwest favored the longhorn because it was resistant to the fever tick. *This page:* The Hereford (top), brought to America in 1817, is now the most popular breed of beef cattle in the United States. Cattle roundups, such as this one (bottom), are a yearly event wherever cattle are allowed free range.

Part of the job of a cowboy is rescuing and returning stray calves to their mothers.

During the winter, western United States beef cattle graze freely on land that may be miles from the actual ranch. Cowboys spend days in the spring rounding up small groups of these cows until the entire herd is once more together. *Below:* Feeding cattle at troughs is almost as common now as allowing them to graze freely in pastures.

The last major farm animal to be domesticated was in many ways the most important. Until about 3000 B.C., human mobilit was strictly limited. Any place someone wanted to go, he or sh had to walk; not surprisingly, th world was a restricted place for most people. The horse change all that. What had been just another source of game for hunters suddenly became the means for rapidly expanding th human horizon. In a week or two, a person mounted on a horse could easily cover, not jus a few tens of miles, but *hundred* Whole peoples were suddenly freed of the restraints of distance, and great waves of huma migration swept back and forth across Asia and Europe.

Until it is weaned, a calf will keep fairly close to its mother. ***Below:*** **This Lapp, native of the northern region of Norway and Sweden called Lapland, tends his herd of reindeer. The domestication of reindeer may be almost as old as that of the dog.** ***Opposite:*** **This beef cow represents one of the hundreds of minor breeds of cattle found across the globe.**

The horse was more than a living transport – it was a tool. It could carry loads, particularly when hitched to a wagon, that would strain the resources of even two or three strong men, and it could pull plows and other farm equipment. The horse freed humans from the drudgery of physical labor.

The horses that pulled these wagons and plows were draft animals. Most, such as the Clydesdale and the Percheron, were large, standing close to six feet at the shoulder and weighing up to 2,500 pounds. In the twentieth century, such draft horses are kept mostly for show, since tractors and trucks have taken over their jobs.

Raising riding and racing horses, however, is still a major business. Breeds such as the thoroughbred, the Morgan, and the quarter horse are smaller and faster than draft horses. All of the modern riding breeds are descended from the Arabian, the small, swift horse developed and bred in the deserts of Arabia and North Africa. The first Arabians entered Europe when the Spanish imported them for their wars against the Moors. Later,

***Top to bottom:* In some parts of the world cattle are beasts of burden, as in this photograph of a farmer in India using oxen to plow his field. Oxen, such as these in Egypt, are also used as draft animals. The horns of these oxen in the Maharashtra Province of India resemble those of ancient cattle breeds pictured in Indian and Egyptian paintings. *Opposite:* The brown Swiss ox and other oxen pulled most of the westward-bound wagons in late eighteenth- and early nineteenth-century America.**

The small donkey or burro has been used as a draft animal throughout the world. Here, one is seen pulling a cart in China.

the Spanish would bring these same horses with them to the Americas, which had no native horses. The wild mustang herds of the U.S. Southwest are the descendants of escaped Spanish horses.

Farmers and ranchers still keep the smaller riding horses as work animals. Herding cattle is easier when done from horse-ack, and the best cattle horse is he quarter horse. This quick, nuscular horse is intelligent and works cattle as capably as a good ıerd dog.

Although most farm animals re mammals, birds – particu-arly chickens, turkeys, and geese – are also an important part f the farming economy. These arm birds have supplied every-hing from food (eggs and meat) ɔ insulation (down) to quill pens.

Farmers have raised birds for number of specific purposes ver the centuries. Falcons and awks made effective hunting artners, while the cormorant ecame a living fishing net (a ing around the bird's neck kept from swallowing the fish). igeons were the first depend-ble method of sending messages uickly over long distances.

he mule, a cross between a horse and ɔnkey, is an exceptionally strong, hardy, ıd intelligent animal. American farmers ere still using mules to pull plows and agons well into the 1930's. *Below:* The ng-haired coat of this donkey makes it ell suited to the chilly, damp climate of e Dingle peninsula of Ireland.

Donkeys, despite a reputation to the contrary, are intelligent, affectionate animals. *Below:* This miniature donkey is only two weeks old. *Opposite:* Trucks and tractors have generally replaced draft horses in the United States, but some loggers still use such horses.

Farmers have also domesticated several cold-blooded species. The farming of freshwater fish – trout and catfish, for example – is a growing business in the last quarter of the twentieth century. Even insects are farmed, and both the silkworm and the honeybee are economically important. Silkworms, first cultivated by the Chinese over 3,000 years ago, produce cocoons, each of which is a single thread of silk measuring between 1,000 and 3,000 feet. Beekeeping is another ancient agricultural industry; domestic bees are still important for producing honey and pollinating crops.

The twenty-first century may see new farm animals added to the familiar ones. In Africa, farmers are attempting to domesticate a large antelope called the eland. Not only would this animal be a good source of food and clothing, but it has one great advantage over domestic cattle: it is immune to sleeping sickness (which is transmitted by the tsetse fly). In Canada and Alaska, other farmers are experimenting with raising elk and musk ox. Again both would be excellent sources of meat and would thrive in climates whose winters are too severe for cattle. The farm animal, no matter its species, will continue to play as important a role in the future as it has in the past.

All riding and racing horses, no matter the breed, are descended from the Arabian, the small, swift horse developed in the deserts of Arabia and North Africa.

The riding horse is still popular all over the world, which is why horse farms such as this one near Kent, Connecticut exist.

For many people, farm animals are seen as residents on "Old MacDonald's farm." Legend, folklore, and fiction have imbued these animals with human characteristics, reflecting the companionship that is felt toward them. Farm animals' survival in captivity is often read as "loyalty," and for some, perhaps it is. What else but loyal would you call the animal that answers to the name you give it, or follows you around?

For these reasons, farm animals not only continue to be necessary but continue to touch the heart.

This Croft's Saddler is one of the many breeds of draft horses originally developed in Europe to do the heavy work of pulling wagons and plows. The Clydesdale, with its feathered hoofs, is one of the most famous draft horses in the world. It is still bred in its native Scotland, mainly for show. *Opposite:* Horseshoes protect a horse's hoofs from wear and damage; a horse with a bad or hurt hoof can go lame.

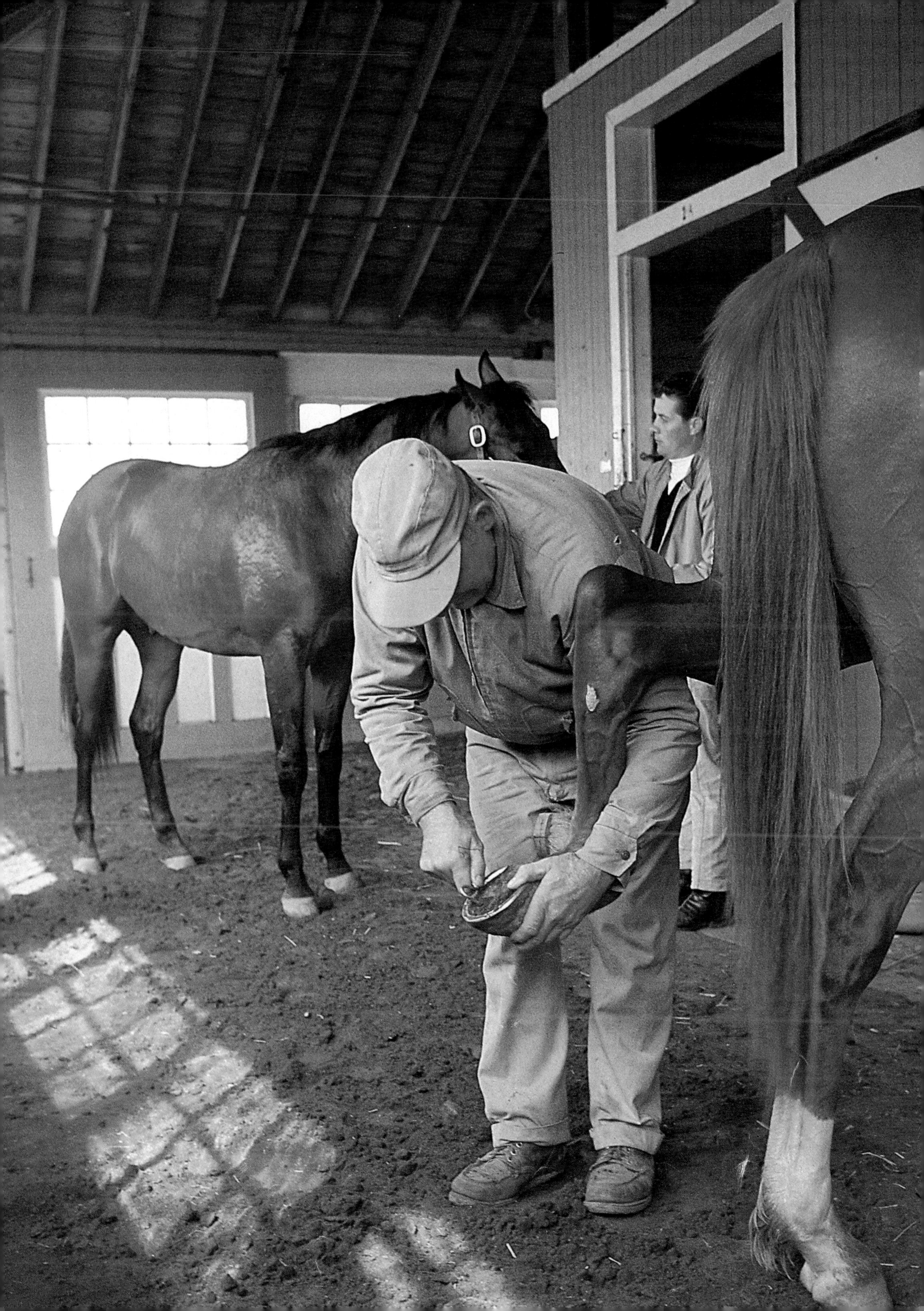

Preceding page: The thoroughbred is the prime racing horse in the world. All horses of this breed are descended from three Arabians imported into Great Britain in the eighteenth century. *This page:* Most horses today (top) are for racing or pleasure riding. Some farms and most ranches, however, still keep work horses, particularly for herding cattle. Ponies, such as this Holland (right), give birth once a year.

Part of a cowboy's job is to break horses for riding. Horses do not take naturally to being ridden, and taming them is both difficult and dangerous. ***Opposite:*** **The quarter horse, originally bred to run quarter-mile races, has been the main mount for cowboys since the last century. It is intelligent and has a natural ability to work cattle.** ***Overleaf:*** **Using his lasso, a cowboy will first rope one of the horses in this corral before he begins saddle-breaking it.**

Index of Photography

All photographs courtesy of The Image Bank except where indicated *

Page Number	Photographer
Title page	Barrie Rokeach
3	Mahaux Photography
4 Top	Peter M. Miller
4 Bottom	Sobel/Klonsky
5 Top	Forrest Smyth/Stockphotos, Inc.*
5 Center	Colin Molyneux
5 Bottom	Harald Sund
6 Top	Yuri Dojc
6 Bottom	Kai U. Mueller/G+J Images
7	Weinberg/Clark
8	Peter M. Miller
9 Top	Walter Bibikow
9 Bottom	Margarette Mead
10 Top	Paul Slaughter
10 Center	Jules Zalon
10 Bottom	Weinberg/Clark
11	Lynn M. Stone
12	Kay Chernush
13	Lynn M. Stone
14 Top	Bruce Wodder
14 Bottom	Alvis Upitis
15 Top	Alvis Upitis
15 Bottom	Lynn M. Stone
16 Top	Paul Slaughter
16 Bottom	Benn Mitchell
17	Alvis Upitis
18-19	Giuliano Colliva
20	David W. Hamilton
21	Lisl Dennis
22	Simon Wilkinson/Stockphotos, Inc.*
23 Top	Joseph B. Brignolo
23 Bottom	Charles Weckler
24	Dag Sundberg
25	Hans Wolf
26 Top	Peter M. Miller
26 Bottom	Nick Nicholson
27 Top	Peter Hendrie
27 Center	Weinberg/Clark
27 Bottom	Tim Bieber
28	Mel DiGiacomo
29 Top	Joe Devenney
29 Bottom	Lawrence Fried
30	Weinberg/Clark
31 Top	Mark E. Gibson*
31 Bottom	Lou Bowman*

Page Number	Photographer
32	Alvis Upitis
33	Yuri Dojc
34 Top	Alvis Upitis
34 Bottom	Sobel/Klonsky
35	Peter M. Miller
36 Top	Joe Azzara
36 Center	Paul Trummer
36 Bottom	Alvis Upitis
37	Peter M. Miller
38 Top	Anne Van Der Vaeren
38 Bottom	Steve Satushek
39 Top	Peter M. Miller
39 Bottom	Anne Van Der Vaeren
40-41	Weinberg/Clark
42	Chuck Place
43 Top	Grant V. Faint
43 Bottom	Cara Moore
44	JaneArt Ltd.
45 Top	JaneArt Ltd.
45 Bottom	Daniel Hummel
46 Top	Lawrence Berman/Stockphotos, Inc.*
46 Bottom	Harold Rose
47	Amanda Clement/Stockphotos, Inc.*
48 Top	Toby Molenaar
48 Center	Lisl Dennis
48 Bottom	Paul Slaughter
49	Lynn M. Stone
50	Guido Alberto Rossi
51 Top	Gill C. Kenny
51 Bottom	John Lewis Stage
52 Top	Lisl Dennis
52 Bottom	Sylvia Schlender*
53	Andrea Pistolesi
54	Douglas S. Henderson
55	Michael Melford
56 Top	Patti McConville
56 Bottom	Erik Leigh Simmons
57	Lawrence Fried
58	Sobel/Klonsky
59 Top	Alexander Schumacher/G+J Images
59 Bottom	Nancy Brown
60	Larry Dale Gordon
60-61	Sobel/Klonsky
62-63	Sobel/Klonsky